# Railroads

*Writer:* T. Harvey
*Designer:* Tri-Art
*Illustrators:* Harry Bishop, Gordon Davies, Roger Full Associates, Tony Gibbons, John Green, Wilfred Harvey, Jack Keay, Wilfred Plowman, Mike Roffe
*Cover Illustrator:* Mike Roffe
*Series Editor:* Christopher Tunney
*Art Director:* Keith Groom

LIBRARY OF CONGRESS CATALOGING IN PUBLICATION DATA

**Harvey, T.**
Railroads.

(The Question and answer books)
First published in 1978 under title: The romance of the railways.
Includes index.
SUMMARY: Questions and answers present the history, construction, and operation of railroads and discuss the various types of trains and engines.

1. Railroads—Juvenile literature. [1. Railroads. 2. Questions and answers] I. Bishop, Harry. II. Title.

TK148.H35    1980              625.1           79-5062
ISBN 0-8225-1184-3    lib. bdg.

This revised edition © 1980 by Lerner Publications Company.
First edition copyright © 1978 by Sackett Publicare Ltd.

All rights reserved. No part of this publication may be reproduced, stored in a retrieval system, or transmitted in any form or by any means, electronic, mechanical, photocopying, recording, or otherwise, without the prior written permission of the Publisher except for the inclusion of brief quotations in an acknowledged review.

International Standard Book Number: 0-8225-1184-3
Library of Congress Catalog Card Number: 79-5062

Manufactured in the United States of America

2  3  4  5  6  7  8  9  10  85  84

# The Question and Answer Books
# Railroads

- 4 Early Railroads
- 6 Railroads across the World
- 8 Making the Passenger Comfortable
- 10 The Great Trains
- 12 Carrying the Freight
- 14 The Permanent Way
- 16 Railroad Engineering
- 18 Steam Engines
- 20 Electric and Diesel Power
- 22 Strange Trains
- 24 Controlling the Trains
- 26 Mountain and Narrow Gauge Railroads
- 28 Underground and Rapid Transit Trains
- 30 Museum Pieces
- 32 A-Z of Railroads
- 34 Wheel Arrangements
- 35 Index

Lerner Publications Company • Minneapolis

# Why do we have railroads?

**EARLY RAILROADS** The idea of laying down a track for vehicles to run along is at least 4,000 years old. But the first real railroads were the mine tracks of central Europe and the wagonways of England. Early public railroads carried freight. Later, they carried passengers, too. Railroads are called by many names. In Britain, the term "railway" is commonly used. *Wagonway* refers to the early timber tracks, and *tramway* to the systems using L-shaped rails.

Railroad tracks make it possible for a single, powered vehicle—the *locomotive*—to pull several other vehicles—the *cars* or *carriages*—easily and safely. The tracks guide the wheels of the vehicles passing over them. Because the tracks are strong and are built on carefully prepared foundations, they can carry very heavy loads.

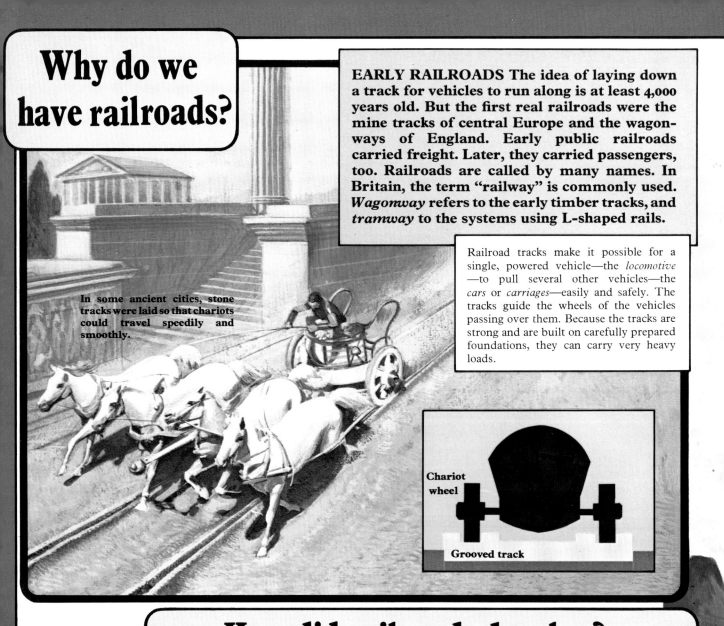

In some ancient cities, stone tracks were laid so that chariots could travel speedily and smoothly.

Chariot wheel

Grooved track

# How did railroads develop?

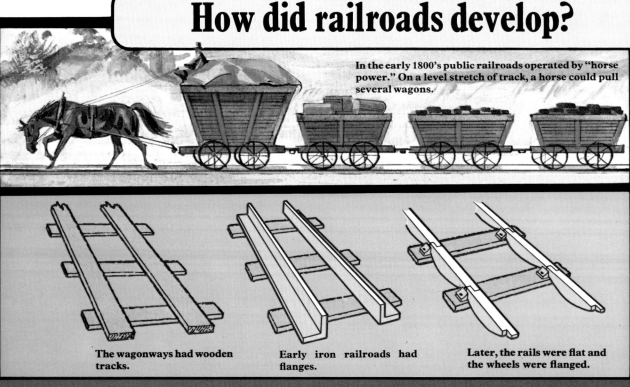

In the early 1800's public railroads operated by "horse power." On a level stretch of track, a horse could pull several wagons.

The wagonways had wooden tracks.

Early iron railroads had flanges.

Later, the rails were flat and the wheels were flanged.

# What was rail travel like in the mid-1800's?

The first passenger cars looked very much like stage coaches mounted on train wheels. Sometimes, several were joined together on one chassis. Fares varied according to the comfort offered. The cheapest cars were merely open trucks. But even in the expensive first-class cars, there was no heating, lighting was poor, and no refreshments were available.

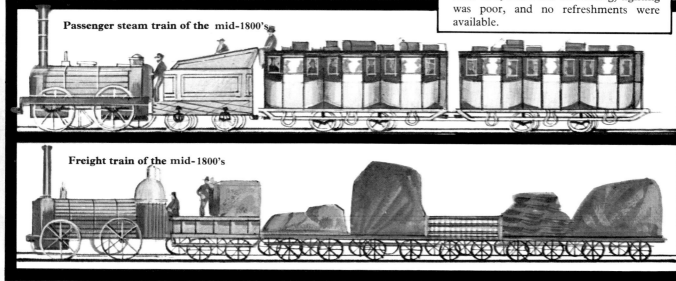

Passenger steam train of the mid-1800's

Freight train of the mid-1800's

# How were the first railroads built?

Most of the railroads of Europe and North America were built by the toil of thousands of laborers. Their tools were picks, shovels, and wheelbarrows. None of the earth-moving machinery of today was available. The gangs of railroad builders moved from job to job. They lived a hard, rowdy life in shanty towns of huts and tents.

## When did the railroad first cross America?

**RAILROADS ACROSS THE WORLD** Once railroads had proved successful, the new form of transportation quickly spread throughout the world. Many of the locomotives were constructed in Britain and the United States. In most of the continents—apart from Europe and North America—railroads were first constructed in the coastal regions. Some lines were built into the interior to help transport natural resources.

Railroads played a major part in opening up the remoter regions of the western and southern United States. May 10, 1869 is an historic date in American railroad history. On that day, tracks from east and west were joined together near Great Salt Lake, Utah, completing a railroad route across North America.

The "iron horse" was sometimes given a hostile reception in its early days.

## Do all railroads carry passengers?

The Swaziland Railway was built to transport ore from rich iron deposits.

As the railroads developed, passenger traffic became increasingly important. However, in most countries, it has now been declining for many years. Today, new railroads are often built just to move heavy bulk loads, and in some countries all railroads are for freight only. In Australia and Africa, most of the new lines are built for freight traffic. But in Africa, passenger trains remain important on the systems of the east and south.

# Where are the world's highest railroads?

All of the 10 highest railroad summits in the world are in South America. Among them, the railroads of Peru and Bolivia reach the greatest altitudes. The first railroads in South America were built along the coast, and were used for carrying freight. The continent has many different track gauges (widths), and electric, diesel, and steam locomotives are all in use.

A railroad station in the Andes. Trains run through the deep Urubamba canyon, near Cusco in Peru.

# What are the older Asian railroads like?

A train runs through a city street in eastern India.

Asia has many types of railroads. They range from the high-speed Lightning trains of Japan to the very old and very slow trains on the narrow gauge lines of the Himalayas. By the standards of industrial countries, many of the chief trains are very old-fashioned. The cars are made of wood, and mixed trains—of freight and passengers—are common.

# Where is the longest straight stretch of track?

It was not until 1970 that the first standard-gauge train crossed Australia from Perth to Sydney—a journey of 2,461 miles (3,960 km). This route includes the longest straight stretch of track in the world, 297 miles (478 km) across the Nullarbor Plain. The most luxurious train on this service is the Indian Pacific Express.

# What comforts can today's passengers expect?

**MAKING THE PASSENGER COMFORTABLE** The first fare-paying passengers in the world were carried in 1807 on the Swansea and Mumbles Railway in South Wales. In the United States, the Baltimore and Ohio Railroad became the first line to offer a public scheduled service.

Club bar on the Indian Pacific in Australia

Dining car of a luxury express in France

Some trains provide passenger cloakrooms.

Sleeping compartments can be spacious and comfortable.

Modern train cars are designed for relaxing travel.

Today's mainline passenger trains offer the traveler many services. Most have dining cars and bars. Air-conditioned coaches are becoming common, and some trains provide secretarial services and telephones. However, passenger traffic has declined in many countries, partly because more and more people have automobiles.

# Do passenger services pay?

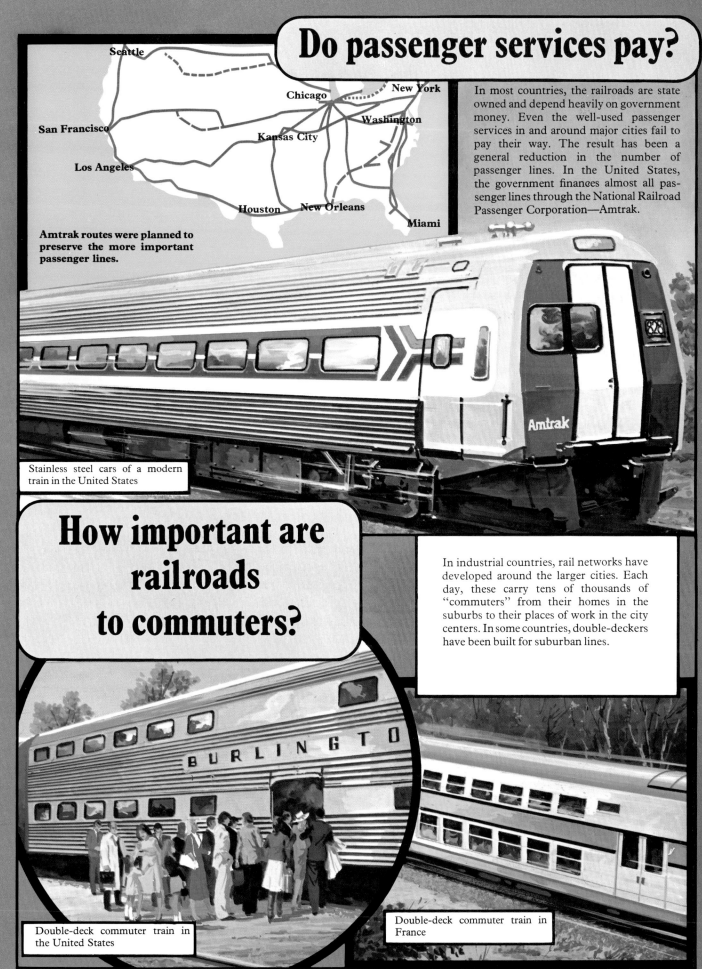

Amtrak routes were planned to preserve the more important passenger lines.

In most countries, the railroads are state owned and depend heavily on government money. Even the well-used passenger services in and around major cities fail to pay their way. The result has been a general reduction in the number of passenger lines. In the United States, the government finances almost all passenger lines through the National Railroad Passenger Corporation—Amtrak.

Stainless steel cars of a modern train in the United States

# How important are railroads to commuters?

In industrial countries, rail networks have developed around the larger cities. Each day, these carry tens of thousands of "commuters" from their homes in the suburbs to their places of work in the city centers. In some countries, double-deckers have been built for suburban lines.

Double-deck commuter train in the United States

Double-deck commuter train in France

# Which is the longest train journey?

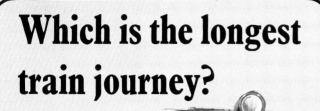

THE GREAT TRAINS The main inter-city and transcontinental trains quickly gained a reputation for speed and comfort. Many of them had romantic names, such as *The Orient Express* and *The Blue Train*. Their fame increased as they gave inspiration to writers, artists, film makers, and composers. Luxurious cars built by the Pullman Car Company were attached to many trains—at first in the United States, and later in several other countries too.

Russian dining cars are often pleasant and inviting.

The Trans-Siberian Express connects Russia's capital city, Moscow, with the port of Vladivostok on the Pacific Ocean.

The longest railroad route in the world connects Moscow with Vladivostok in Russia. Known as the *Trans-Siberian* route, it is 5,801 miles (9,336 km) long. The train that uses the railroad is named *The Russia*, and it sets out daily on its eight-day journey. It carries local people as well as long-distance travelers.

# What was the Orient Express like?

One of the most famous of all long-distance trains was *The Orient Express*, which ran between Paris and Istanbul (formerly Constantinople). It made its first journey in 1883. Travel on *The Orient Express* was considered adventure. And it had an air of romance and excitement that made it a favorite with thriller writers.

The route of the Orient Express ran through 11 countries.

The California Zephyr skirts a mountain river.

# Which is the most beautiful route?

Railroads have been built in some of the world's most beautiful places. And in most continents, magnificent scenery can be seen from train windows. But many people have claimed pride of place for a famous American train, the *California Zephyr*, with its route across the spectacular landscapes of Colorado.

# How many countries are crossed by the Trans-Europ-Express?

# How fast are the Hikari trains?

The luxurious TEE links several European countries.

Observation cars give TEE passengers an all-round view.

The *Lightning* (*Hikari*) trains provide a high-speed service between Tokyo and Osaka in Japan. They cover the 320 miles (515 km) in just over 3 hours. Each train has seating for 1,400 passengers. Japan's new railroad system will allow speeds up to about 160 mph (260 kph).

The Lightning trains are smooth, fast, and efficient.

The *TEE* or *Trans-Europ-Express* was born in 1954. It is a system of express trains administered from The Netherlands. Its 28 routes cross Belgium, France, The Netherlands, West Germany, Italy, Luxembourg, Switzerland, Austria, and Spain. One of its famous trains is the *Rheingold*, between the Hook of Holland and Geneva.

## Are there special cars for special loads?

**CARRYING THE FREIGHT** Freight-carrying was the main reason for the building of the first railroads. Loads were generally moved short distances only, for example from mine to canal, or from canal to factory. And railroad operators thought in terms of "car loads." Today, railroads often operate on the basis of "train loads," and freight trains frequently travel directly between mine and factory or between factory and factory.

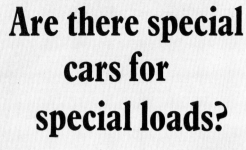
Car for bulky goods

Vehicle transporter

Car for carrying rolls of sheet metal

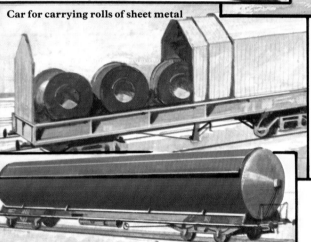
Agricultural-produce car

Liquid gas tanker

Double car for carrying very heavy equipment

A whole range of railroad cars has been built to cope with the many different loads carried by the railroads. In designing new cars, particular attention is paid to ease of loading and unloading—especially in the case of cars carrying bulk materials such as iron ore, coal, powders, and liquids.

## How are cars sorted?

Freight yards sort cars according to destination.

Cars arrive — Hump — Cars sorted — Trains depart

From the "hump," each car coasts to join its train.

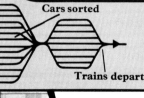
The control center switches cars to the correct tracks.

Cars are sorted and formed into trains in freight yards. Some such yards are computer-controlled. Each car is pushed to a hump, and then automatically directed to run into a siding, where the various cars of its train are being coupled together.

# What is container traffic?

Container handling is a means of transporting goods without the need for constant loading and unloading. Goods are placed in large, box-like containers at their point of origin, and are not removed until they reach their final destination. During the journey, the container may be carried on a road vehicle, a train, or a ship, or even in an aircraft.

A heavy crane transfers a container from a train to a truck.

# Are there freight expresses?

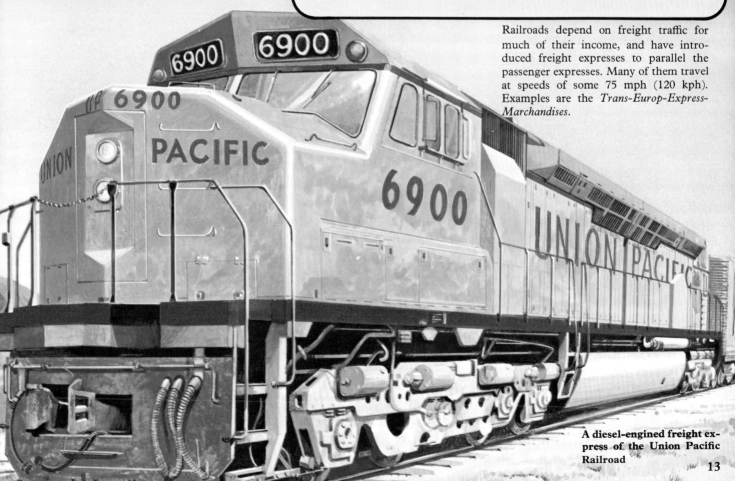

Railroads depend on freight traffic for much of their income, and have introduced freight expresses to parallel the passenger expresses. Many of them travel at speeds of some 75 mph (120 kph). Examples are the *Trans-Europ-Express-Marchandises*.

A diesel-engined freight express of the Union Pacific Railroad

# How is train track laid?

In the past, laying train track was slow and difficult.

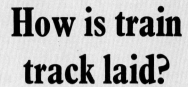

Modern track-laying machines are fast and efficient.

**THE PERMANENT WAY** Keeping the train track in good order is essential to the safe running of the trains. Track maintenance is expensive, and the railroads are always seeking ways of reducing maintenance costs without affecting safety. Much of the construction work is now done by machines, and is no longer the back-breaking task it once was. And the use of long welded rails results in less maintenance at the rail joints, and a smoother ride.

The railroad track, or *permanent way*, consists of three main parts. First, there are the rolled-steel rails on which the trains run. Second, there are supporting ties made of wood or concrete, though steel is used in some countries. Third, there is the packing around the ties which prevents movement and spreads the weight of the trains.

# How is the track held down?

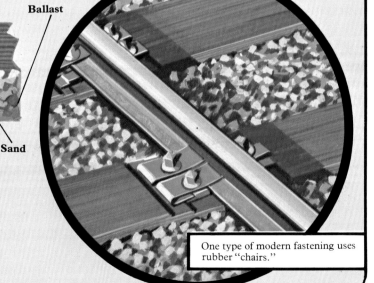

Originally, the rails were held onto the ties by "chairs." To hold the rails firmly in the chairs, wooden wedges were used. After the introduction of flat-bottom rails, it became possible to fasten the rails directly onto the ties. One of the most successful rail fastenings is the "Pandrol" spring-steel clip.

One type of modern fastening uses rubber "chairs."

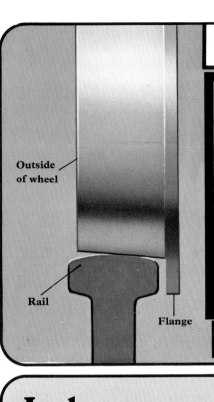

*Left:* Flanged wheels hold trains on the track.

*Above:* Some early railroads used flanged rails.

# Why does a train stay on the track?

A train stays on the track because its wheels are flanged—they have a protruding rim that prevents them from moving sideways. Today, wheels are almost always flanged on the inside. But many kinds of track and wheel arrangements have been tried. Early railroads used flanged rails in preference to flanged wheels. On modern express trains, the suspension is so good that a wheel flange rarely touches a rail.

# Is there a standard track width?

Track gauges (widths) vary in different parts of the world. The most common standard gauge is 4 ft $8\frac{1}{2}$ inches (1.435 meters). It is used in North America, Australia, Japan, most of Europe, and many other places. Some countries, including Australia and Japan, use more than one gauge.

**Where two systems meet, the track may have two gauges.**

# How do switches work?

*Below left:* The switch moves to direct trains right.

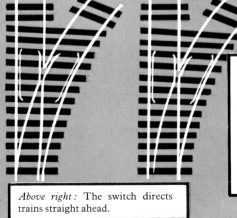

*Above right:* The switch directs trains straight ahead.

The switch may be very simple—as where just one siding leaves the main line—or very complicated—as where many lines cross and join outside major stations. Originally, all switches were controlled manually by means of rods that went from the signal tower to the movable parts of a switch. On most main lines, switches are now operated electrically by just pushing a button.

# Where is the longest railroad tunnel?

**RAILROAD ENGINEERING** Engineers have to consider many factors when planning a new railroad. First, there is the question of gauge—the line must be able to link up with those nearby. Then, bridges and viaducts must be strong enough to bear the weight of trains. And curves in the line must not be so sharp that they will cut down speed. And, above all, there is the problem of "flattening out" the countryside so that the line is level.

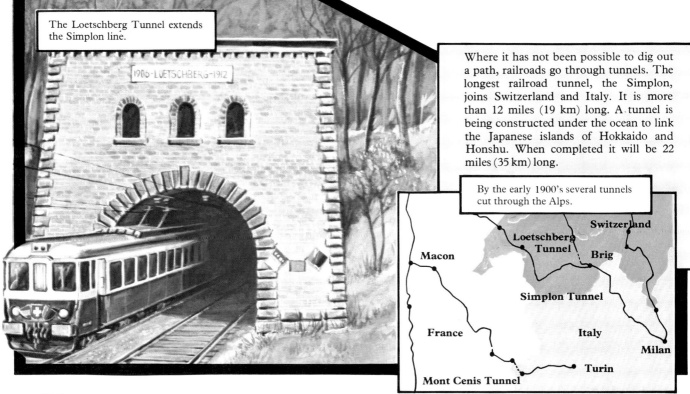

The Loetschberg Tunnel extends the Simplon line.

Where it has not been possible to dig out a path, railroads go through tunnels. The longest railroad tunnel, the Simplon, joins Switzerland and Italy. It is more than 12 miles (19 km) long. A tunnel is being constructed under the ocean to link the Japanese islands of Hokkaido and Honshu. When completed it will be 22 miles (35 km) long.

By the early 1900's several tunnels cut through the Alps.

# When were the first major train bridges built?

Viaducts and bridges are expensive and difficult to build but, often, they cannot be avoided when constructing a railroad. In the late 1800's, major railroad bridges were already being built. One of the most famous of them was the great cantilever built across the Firth of Forth in Scotland. Begun in 1882, it was opened in 1890.

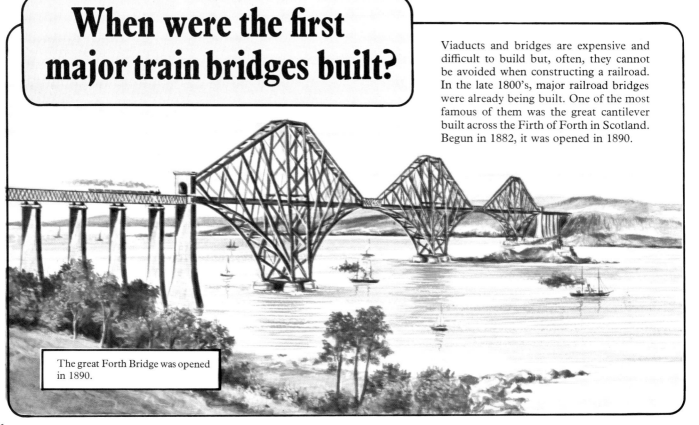

The great Forth Bridge was opened in 1890.

# How is the ground "smoothed out" to build a railroad?

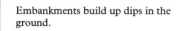

Embankments build up dips in the ground.

Paths are driven through small hills on railroad routes.

Trains operate most efficiently on a level track. The railroad engineer chooses routes that are as level as possible. Even so, it is often necessary to undertake major earth-moving operations. *Paths* are dug through small hills. The earth from a cutting can often be used to build up an *embankment* where the line crosses a dip in the ground.

Viaducts carry railroads over valleys and rivers.

# How do railroads climb hills?

Train tracks curve from side to side to climb hills.

In mountainous regions, it may be necessary for a railroad line to be built in spirals in order to climb a hill. One line in Switzerland has a series of four spirals in tunnels. Two spirals help the Canadian Pacific line over the Rocky Mountains.

# What are modern train stations like?

Modern train stations are bright and efficient.

Like the railroads they serve, stations are built in all shapes and sizes, from a single small wooden platform to the vast Grand Central in New York City. The large modern stations closely resemble airports. They often have shops as well as the usual facilities for passengers, such as rest rooms, bars, and restaurants.

Many older stations look like "glass cathedrals."

# How does a steam engine work?

**STEAM ENGINES** The early railroads used trains drawn by horses, but the development of the steam locomotive was the most important event in railroad history. For more than a hundred years, steam engines dominated the railroads of the world. Though they are still used in many countries, they have been entirely replaced in many others by the more efficient—but less romantic—diesel and electric locomotives.

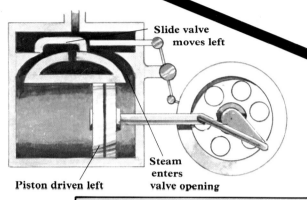

*Below right:* The slide moves, and directs steam to the other side. The piston is driven back to the left.

*Above:* Steam from the boiler enters the cylinder and drives the piston to the right. Connecting rods turn the wheel.

In a steam engine, water is heated by a fire and changed to steam. As the steam expands, it drives pistons backward and forward inside cylinders. The pistons are linked to the driving wheels of the locomotive, and their movements cause the wheels to turn.

# What is a steam locomotive like inside?

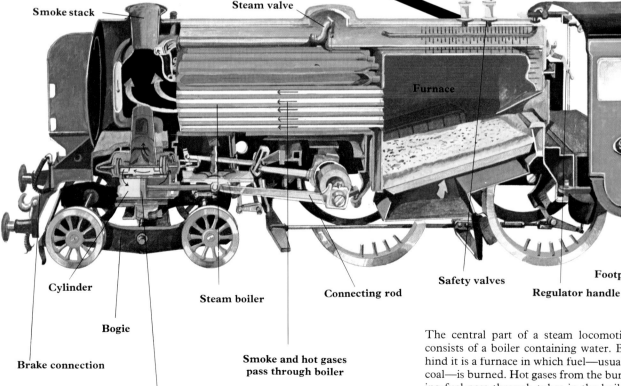

The central part of a steam locomotive consists of a boiler containing water. Behind it is a furnace in which fuel—usually coal—is burned. Hot gases from the burning fuel pass through tubes in the boiler, heat the water, and produce steam. The steam is superheated—made even hotter—before being piped to the cylinders.

# How has the locomotive developed?

Robert Stephenson's Rocket (1829)

Richard Trevithick's locomotive (1804)

The American locomotive Best Friend (1830)

Classic American locomotive of the mid-1800's

During the many years of locomotive building, the general trend has been towards increased power, size, and weight. But all steam locomotives operate on the same principle—the principle that was established by Robert Stephenson when he built his famous *Rocket* in 1829. By the 1930's, steam locomotives had reached their peak of efficiency.

The great days of steam: a giant locomotive of the mid-1900's

The post-steam era: a modern diesel locomotive

# How long have electric and diesel locomotives been in use?

**ELECTRIC AND DIESEL POWER** Although steam locomotives are still used in some parts of the world, most industrial countries now use diesel and electric locomotives. The most efficient locomotive of all is the electric, but converting a railroad to electric traction is very costly. However, most of the new high-speed trains are powered by electricity.

Although diesel engines had been invented and used in locomotives before the end of the 1800's they did not become common until the 1920's. The first electric train ran at the Berlin Trade Exhibition in 1879. Several public electric railroads were in use by the 1800's

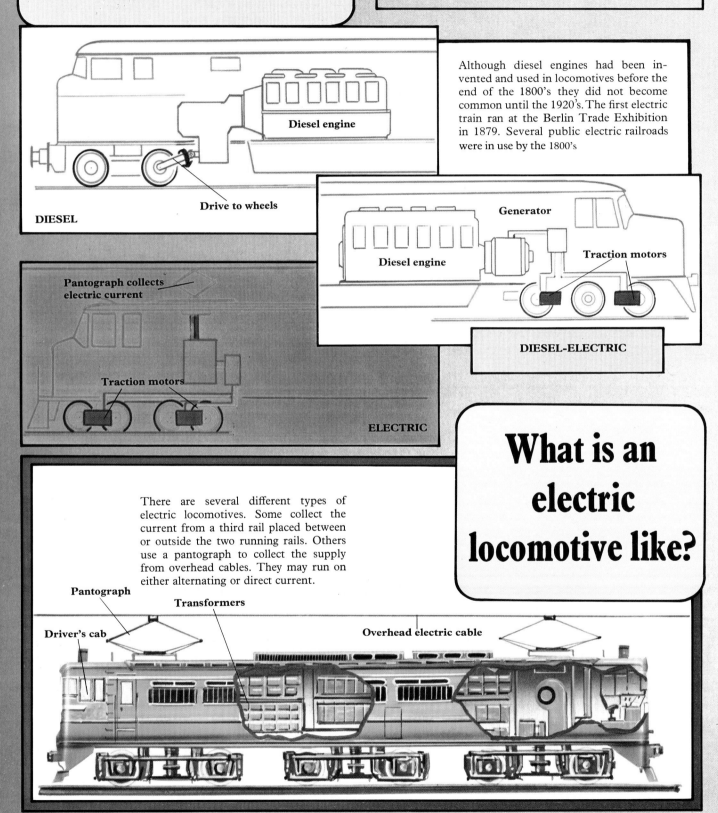

# What is an electric locomotive like?

There are several different types of electric locomotives. Some collect the current from a third rail placed between or outside the two running rails. Others use a pantograph to collect the supply from overhead cables. They may run on either alternating or direct current.

# What advantages do diesel locomotives have?

Diesel locomotives have many advantages over steam locomotives. They are cleaner, are more efficient in their use of fuel, and need only one person to drive them. In addition, they are easily prepared for a journey, and can cover longer distances in a day than steam engines.

**Driver's cab** **Air intake**

**Traction motors** **Generator** **Diesel engine**

# Diesel-Electric Locomotives

The most common type of diesel locomotive is the diesel-electric. Its main parts are a diesel engine, a generator, and a number of electric motors. The engine turns the generator to produce electricity. This is then fed to electric motors that drive the wheels. The motors are mounted on the bogies.

**Turbine**

**Rotary compressors**

**Air intake** **Exhaust**

**Fuel tanks**

**Traction motors** **Generators** **Transmission**

# Strange Trains

A screw ran the whole length of the railroad, mounted in a channel between the rails. The thread of the screw engaged with a projection under each train car. An engine turned the screw, and the screw pulled the train along.

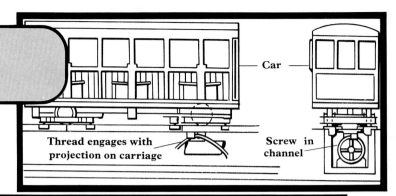

**Thread engages with projection on carriage** — **Screw in channel** — Car

### THE SCREW-DRIVEN TRAIN

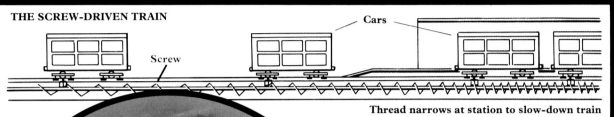

Cars — Screw

**Thread narrows at station to slow-down train**

### LARTIQUE MONORAIL

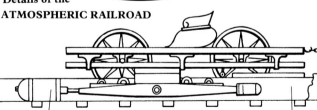

Train balanced on raised rail

### DOUBLE ENGINE

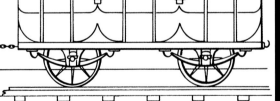

The double engine was invented by Robert F. Fairlie in the 1860's. It was really two engines in one, and as a result was very powerful.

### Details of the ATMOSPHERIC RAILROAD

Piston attached to train — Airtight tube

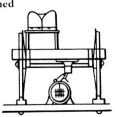 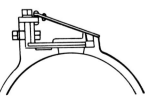 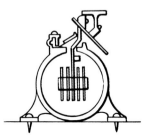

An airtight tube was mounted between the rails. A piston, driven along the tube by air pressure, pulled the train with it.

# Trains of the Future?

Engineers try constantly to improve railroad efficiency. Their major considerations are speed, safety, reliability, smoothness, quietness, and freedom from pollution. When new types of railroad vehicles are invented, they are tested to see whether they really are improvements. Recent experimental vehicles include the German MBB and the British tracked hovercraft.

The **MBB (Messerschmitt-Bolkow-Blohm) has no wheels. It is held in the air by magnetic force.** Magnets attached to the vehicle "float" along under the broad rails beside the track.

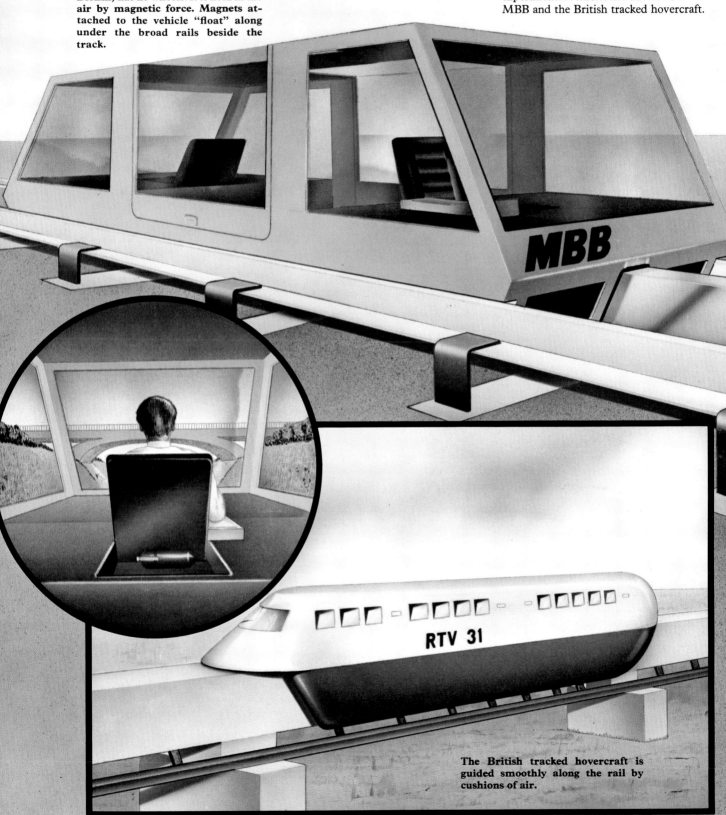

The **British tracked hovercraft is guided smoothly along the rail by cushions of air.**

# What do railroad signals mean?

**CONTROLLING THE TRAINS** Safety is even more important than speed and comfort. One of the chief concerns of railroad operators is the improvement of train control. The trend is toward the increasing use of computers and automation, and even many level crossing gates now work automatically.

At night, the driver looks for the colored light on the semaphore arm—red, amber, or green.

Semaphore signal meaning "Stop"

Semaphore signal meaning "Proceed with caution"

Semaphore signal meaning "Proceed at full speed"

The driver of a train must know what is happening on the track ahead. That is the purpose of signals. The earliest signals consisted of people waving flags. Later, signals were mounted on posts, but were still controlled individually. Still later came the semaphore signals that are still in use on many sections of track today. Increasingly, semaphores are being replaced by colored-light signals.

Some signaling systems use lights only.

# What is a signal tower like inside?

In the past, signals were operated by heavy levers.

Today, many signal systems are electronically controlled.

In the early days, each signal tower controlled only a short section of track. Now, one signal tower can control hundreds of miles of line. Large levers in the old signal towers were used to pull and push the wires and rods that operated signals and switches. The signal operator of today routes trains over complex tracks at the push of a switch.

# What is the driver's cab like in a modern train?

The driver's cab is neat and efficient.

Controls and instruments are clear and simple.

The modern train driver's cab is rather like that of a truck. It is very different from the "footplate" of the old steam locomotive with its dirt, smoke, and heat. Steam engines needed much preparation before they were ready for the day's work, whereas a diesel or electric engine is ready to go at the flick of a switch. However, despite all the modern aids, the train driver of today still has a demanding and responsible job.

The "footplate" was romantic but dirty.

# Which is the steepest rack railroad?

**MOUNTAIN AND NARROW GAUGE RAILROADS** Many mountain railroads have narrow gauge tracks because sharper curves are possible on narrow gauge than on standard gauge. In addition, narrow gauge railroads are cheaper to build, and many mountain railroads are used only by tourists. Where the climb is particularly steep, the rack and pinion system is often used. It prevents any possibility of the wheels slipping on the track. Narrow gauge railroads also have industrial uses—in factory yards, for example.

The Mt Pilatus Railroad is the world's steepest. It climbs from Lake Lucerne to the Pilatus summit.

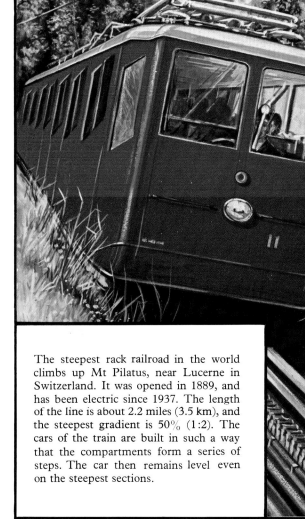

# How does the rack and pinion system work?

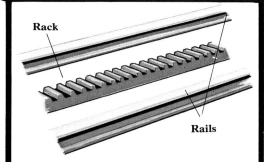

John Blenkinsop (1783–1831) doubted that an engine with smooth wheels could grip a smooth track. He therefore designed a system of toothed wheels running along a rack. Although his idea proved unnecessary for the majority of railroad lines, it is the ancestor of the rack-and-pinion systems used today on mountain railroads. Most of them are based on the railroad ideas of Roman Abt (1850–1933), in which a minimum of two racks are fixed between the running rails. These have teeth in their upper edge that lock into pinions (toothed wheels) on the underside of the locomotive.

The steepest rack railroad in the world climbs up Mt Pilatus, near Lucerne in Switzerland. It was opened in 1889, and has been electric since 1937. The length of the line is about 2.2 miles (3.5 km), and the steepest gradient is 50% (1:2). The cars of the train are built in such a way that the compartments form a series of steps. The car then remains level even on the steepest sections.

# Which railroad goes through a famous mountain?

The rack and pinion Jungfrau Railroad climbs for about 6 miles (9 km) up into the Alps.

The north face of the Eiger Mountain in Switzerland has claimed the lives of many climbers. The less daring can, however, ascend it in comfort by using the train from Kleine Scheidegg to Jungfraujoch. The railroad, which has used electric traction since 1912, has an average gradient of 1:4. It reaches a height of 11,332 feet (3,454 meters) at its summit. It is the highest railroad in Europe.

# Where is the world's smallest public railroad?

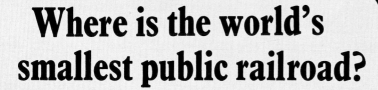

The Romney, Hythe and Dymchurch is the smallest public railroad in the world. Its locomotives are one-third full size.

A well-known racing driver, Captain J. E. P. Howey, built the tiny Romney, Hythe and Dymchurch Railway to link the stations of Hythe and New Romney in south-east England. It was constructed as a public railroad, but to a gauge of only 15 inches (38 cm). The line was opened in 1927, and was extended to Dungeness in 1929. Its total length is about 14 miles (22 km). It has 10 miniature locomotives, which are modeled after famous mainline engines.

# How do railroads aid city communications?

**UNDERGROUND AND RAPID TRANSIT TRAINS** In most cities, travel has become very difficult because of the increasing number of cars on the roads and the increasing number of people wanting to make journeys. To overcome this problem, many cities have built railroads that allow the rapid movement of large numbers of passengers. Most of these are underground systems and highly automated suburban railroads.

Several large cities embarked on the building of underground railroads in the early 1900's They hoped to relieve congestion on the roads. Many countries are now extending their underground systems, and building new ones. One completely new railroad system, which is almost entirely automated, is the Bay Area Rapid Transit of the San Francisco-Oakland area of California.

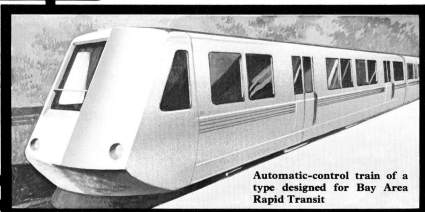

**Automatic-control train of a type designed for Bay Area Rapid Transit**

**Shuttle train in Switzerland**

**Overhead monorail in Japan**

# Which was the first underground railroad?

The first underground railroad was opened in London in 1863. It was steam operated, and must have been very dirty to travel on. The first electric underground railroad was also in London, and opened in 1890. It was followed 10 years later by the Paris *Métro*.

**A Gooch steam engine pulling a train on London's Metropolitan Railway—the world's first underground railroad.**

**Modern train of the London underground**

# Which is the most beautiful underground railroad?

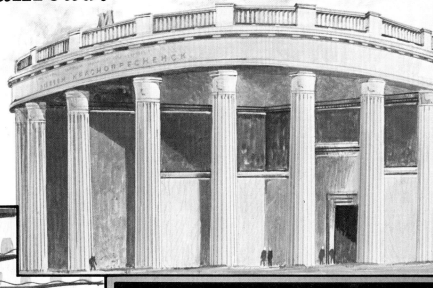

*Entrance to a station of the Moscow underground*

The spacious marble halls of the Moscow underground are known all over the world. The Moscow underground was built in 1935, and carries about 1.8 billion passengers each year. At present, it has about 100 miles (160 km) of track, but eventually this will be doubled. Many other Russian cities also have underground trains. They include Baku, Kharkov, Kiev, Leningrad, Tashkent, and Tbilisi.

*Interior of a Moscow underground station*

## The Vital Underground

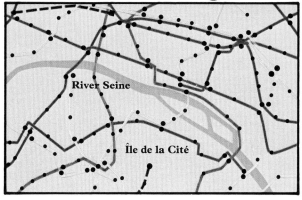

Underground trains are useful in cities because they know no obstacles and can even pass under rivers. This map of the *Métro* shows how its lines reach all parts of Paris.

# Are trams still used?

Many cities have long abandoned their tram systems. But others are making them the basis of their new rapid transit services. Tramlines are being modernized and extended to the suburbs. In city centers, they are put underground. To cater to the large number of travelers, several cars can be linked together to form a "train."

*Articulated tram in the Isle of Man—a country tram*

# Have many railroads been closed?

MUSEUM PIECES In the last 30 years, railroads have suffered from competition by airlines and road vehicles. As a result, many lines have been closed. With the loss of so many railroads, it has become important to preserve relics of the past railroad age. They form a significant part of industrial and social history.

Railroads are shrinking all over the world. In 1916, the United States reached its maximum track length—254,000 miles (408,800 km). By 1972, this had been reduced to 208,998 miles (336,350 km). In some European countries, an even greater proportion of track has been lost.

In many places, ruined stations and overgrown track are all that is left of once-busy railroad lines.

# How do people preserve old trains?

The most dramatic and satisfying method of preservation is actually to run a railroad. Most such preserved railroads use steam engines. Railroad preservation societies are found world-wide. In Europe alone there are hundreds, and in the United States there are more than 80.

A working replica of George Stephenson's Locomotion (1825)

Many societies preserve great engines of the early 1900's.

# Where is the world's fastest steam engine?

On July 3, 1938, the London and North Eastern Railway's A4 streamlined locomotive *Mallard* reached 126 mph (202.8 kph), hauling a train of 240 tons. It is now preserved in the National Railway Museum at York, in England. Many countries have founded national museums to illustrate the history of their transport systems, and there are numerous smaller privately owned displays.

When the *Mallard* went into service, the steam age was drawing to a close. Soon, the development of steam locomotion ceased.

# Model Railroads

Many famous locomotives and trains of the past are remembered today chiefly by model-railroad enthusiasts. Some models are very accurate and detailed.

# A-Z of Railroads

## A

**adhesion** The grip between train wheels and the rail. Early engineers thought that because both the wheels and the rail had smooth surfaces, a train would not have enough grip to move at speed. This fear proved groundless.

**air brake** Train brakes operated by compressed air. Brakes may also be operated hydraulically or electrically.

**ATC** Automatic Train Control has been tried in several countries, and varies from very simple systems to fully automated trains that can operate without a driver.

**AWS** Automatic Warning System provides both visual and audible warnings to a train driver of signals he has passed.

## B

**ballast** The crushed rock, often granite, that is put between the ties of a railroad track. It prevents movement of the track, and helps to distribute the weight of the trains.

**banking** The attachment of a second locomotive to the rear of a train to provide assistance up steep inclines. It was common in the days of steam.

**Beyer-Garratt locomotive** A special type of steam locomotive consisting of two separate engines. The most famous Beyer-Garratts were used in southern and eastern Africa. They were built by the firm of Beyer Peacock.

**Big-Boy** The largest steam locomotives ever built. They were used on the Union Pacific Railroad in the United States. Their overall length was 132 ft $9\frac{3}{4}$ inches (40.48 meters).

**Black Five** A famous class of 4-6-0 steam locomotives used for hauling freight and passenger trains on the London Midland and Scottish Railway in Britain.

**block system** The division of the line by signals into sections. Only one train at a time is allowed into a section.

**boiler** The component of a steam locomotive in which water is heated to make steam. The main parts of a boiler are the furnace, the barrel, and the smoke box.

**branch line** A line separate from the main line, connecting small towns and villages. Branch lines are often short, and, in several countries, many have been closed in recent years.

**broad gauge** A railroad line gauge greater than standard gauge. *See* STANDARD GAUGE.

**buffet car** A coach in which light meals and drinks are served.

**bulk load** Freight that is transported loose—for example, grain, cement, and coal.

## C

**cab** The part of a steam locomotive in which the engineer and fireman work. It contains the controls and various gauges, as well as the furnace doors through which fuel is thrown on the fire.

**caboose** The car at the end of a freight train. It is used by the conductor.

**car** A vehicle for carrying passengers. Many modern train cars are open plan, but the older cars are generally divided into a number of compartments.

**Catch-me-who-can** A locomotive built by Richard Trevithick (1771–1833) and demonstrated on a circular track in London in 1808.

**chair** A metal bracket to which the older types of rails were fixed. The chair was fixed to a tie.

**classification yard or control yard** The specially equipped place in which freight cars are sorted and then reassembled into trains so as to reach their correct destinations.

**communication cord** A cord running the length of a train that passengers can pull in an emergency to stop the train. Modern trains often have an emergency warning or braking system operated by handles in each car.

**commuter** A person who daily travels to work from his home either by train or some other means of transportation.

**compartment** *See* CAR.

**container traffic** The use of containers —large metal boxes—for the transfer of freight.

**continuous welded rail** Railroad track formed of short lengths of rail welded together. Continuous welded rail reduces maintenance costs, and also gives passengers a quieter and smoother ride.

**couchette** A train car in which the seats convert into beds. Usually there are six berths to a compartment.

**cutting** An artificial valley cut through a small hill to keep a railroad track level. *See* EMBANKMENT.

## D

**dead-man's handle** A safety device on diesel and electric locomotives. It is a handle that must be held down by the driver while the locomotive is in motion. If the driver lets go of it, power is cut off and the brakes are applied.

**diesel locomotive** A locomotive that uses a diesel engine as its power source.

**diner** A specially built train car in which meals are served. Part of it is arranged as a restaurant, and it may also have a small kitchen.

**distant signal** A yellow-colored semaphore signal that, when set at danger, indicates to a driver that he will have to stop at the next signal.

**dome-coach** A car with a glass observation dome set in the roof. Such cars are common on scenic routes, especially in North America.

**double-decked car** A passenger-carrying car with two levels, usually used for commuter services.

**dynamometer car** A special car attached to a train, containing instruments for measuring the performance of a locomotive.

## E

**electric locomotive** A locomotive powered by an electric motor.
**elevated train** Railroad built on trestles over streets.
**embankment** The opposite of a cutting. The level of the ground is raised artificially to keep railroad track as level as possible.
**erecting shop** The workshop in which steam locomotives are assembled.

## F

**fireman** The second man in the cab of a steam locomotive. He feeds the fire with coal. He is also called a *stoker*. Some steam locomotives have a mechanical stoker.
**flange** The projecting edge of a wheel that keeps the wheel in position on a rail.
**freight** All types of goods carried by train or other means of transport.
**funicular** A cable railroad used in mountainous regions.
**furnace** The rear part of the boiler of a steam locomotive. The coal is burned in it.

## G

**gauge** The distance between the rails of a train track.
**grade crossing** The place where a train track crosses a road on the level. Usually, grade crossings are protected by gates or other barriers, so that only the road or the train track is open at any one time. Formerly the gates at grade crossings were operated by a crossing keeper. Many are now automatically controlled. The term is also used for the place where two sets of train tracks cross each other.
**gradient** The slope of the train track, either upward or downward from the level.
**guard's van** *See* CABOOSE.

## H

**hand signal** Originally all train signals were given by hand. Today hand signals are used only in an emergency.
**home signal** A red semaphore signal that compels a train to stop when set at danger. Home signals are frequently located at the end of station platforms.
**hopper** A freight car that unloads its cargo through its floor.
**horse traction** The use of horses for pulling trains. In the early days of the railroads, both wagons and passenger cars were drawn by horses.

## IJK

**injector** The mechanism in a steam locomotive that forces water into the boiler.
**Intercontainer** An international European organization that promotes container freight handling. More than 20 railroads in various countries cooperate in Intercontainer.
**interlocking** A system of train control in which signals are interlocked with switches. It prevents the signals from being operated independently of the switches they guard.
**joint bar** The metal plate that is bolted to the ends of rails to join them together.
**key** A wedge that holds a rail in a chair. Originally wedges were made of very hard wood, but some more recent ones are of spring steel. *See* CHAIR.

## L

**light engine** A locomotive running on its own without a train.
**light railroad** A railroad built to standards less than those usually required in railroad construction. Usually, trains on such railroads are only permitted to operate at limited speeds.
**loading gauge** The maximum height and width of train cars permitted on a railroad line.

## M

**mail train** A train made up of vehicles to carry letters and parcels—or one that includes some such vehicles among its cars. Some mail trains include cars in which letters are sorted. These are sometimes called *Traveling Post Offices*.
**mainline** A major railroad line linking large cities. Mainline trains usually operate at higher speeds than those on branch lines.
**monorail** A railroad line consisting of a single track. In some monorails, the cars are slung beneath the track. In others they run above it.
**mountain railroad** A railroad in a mountainous region, usually of narrow gauge. In order to climb steep slopes, a rack-and-pinion system may be used. *See* RACK AND PINION.

## N

**narrow gauge** Train tracks with a gauge of less than 3.28 ft (1 meter).

## O

**observation car** A passenger car with large windows to enable travellers to enjoy the scenery. Such cars are often attached to the rear of a train.

## P

**pantograph** The apparatus on top of an electric locomotive that collects current from overhead cables.
**pendulum suspension** A system that enables a train's body to tilt while taking corners.
**piggyback** A term used to describe the conveyance of road vehicles by train. They are driven directly onto and off flat wagons.
**Pullman** The name given to the luxuriously appointed coaches and sleeping cars developed by the American George Mortimer Pullman (1831–97). They were operated mainly on the American railroads although some ran in Europe.
**push-pull** A type of train, often used on branch lines, in which the locomotive is permanently attached to one end, but can be controlled from either end.

## R

**rack and pinion** A system used on mountain railroads to provide adhesion. The locomotives have a cog wheel (the pinion) that fits into a toothed rail (the rack). *See* ADHESION.
**railbus** Small, often four-wheeled, diesel-driven car, of a type sometimes used on branch lines.
**railhead** The end of a railroad line. The term is used especially of a place where arrangements are made to assist the interchange between road and rail of passengers and freight.
**railroad** A transport system by which vehicles are moved along a prepared track. In much wider terms, the word often means the whole system of train operation.
**rails** The strips of steel on which train wheels run.
**rapid transit** A railroad, especially in cities, for conveying passengers quickly over short routes. Many rapid transit systems are underground.
**regulator** The control lever on a steam locomotive.
**Rocket** The famous engine built by Robert Stephenson that won the Rainhill Trials in 1829, and established the principles of steam locomotive design.
**roundhouse** A building in which locomotives are cleaned and repaired. All of the tracks radiate out from a turntable. *See* TURNTABLE.

## S

**sleeper** A train that carries passengers on overnight journeys and provides sleeping accommodation.
**standard gauge** A railroad in which the distance between the rails is 4 ft $8\frac{1}{2}$ inches (1.435 meters).
**station** The place where passengers board a train.
**steam locomotive** A locomotive powered by steam-driven pistons.
**stoker** *See* FIREMAN.
**switch** The means by which a train

moves from one track to another. It involves the use of a movable rail.

## T

**tank locomotive** A locomotive in which a water supply is stored in tanks alongside the boiler, and in which coal is kept in a bunker behind the cab.

**TEE** The *Trans-Europ-Express* trains connect the major cities of Europe. The trains are very comfortable and run at high speed.

**tender locomotive** A locomotive in which the water supply and coal are carried in a separate vehicle, the *tender*, which is coupled to the engine.

**tracks** The railroad line, made up of rails, ties, and ballast.

**train ferries** Ships onto which passenger cars and freight cars can be run directly from the dock. Ferries are usually used for short lake or sea crossings.

**tram** A passenger-carrying car that runs on rails in city streets or on short runs outside cities. Some trams consist of two or more cars linked together.

**tramway** An early railroad that used L-shaped rails. The term is also used for various types of light railroad. *See* LIGHT RAILROAD.

**tube** *See* UNDERGROUND RAILROAD.

**turntable** An apparatus for turning steam engines around. It consists of a short length of train track on a revolving platform.

## UV

**underground railroad** A railroad that runs through tunnels built beneath city streets.

**vacuum brake** *See* AIR BRAKE.

**viaduct** A long bridge, usually crossing low land or a wide valley.

## W

**Wagon-Lit** A coach belonging to the International Sleeping Car Company or *Compagnie Internationale des Wagons-Lits,* founded in 1876.

**wagonways** Early railroads with cars running on wooden tracks. Sometimes the word is spelled *waggonway*.

## Wheel Arrangements

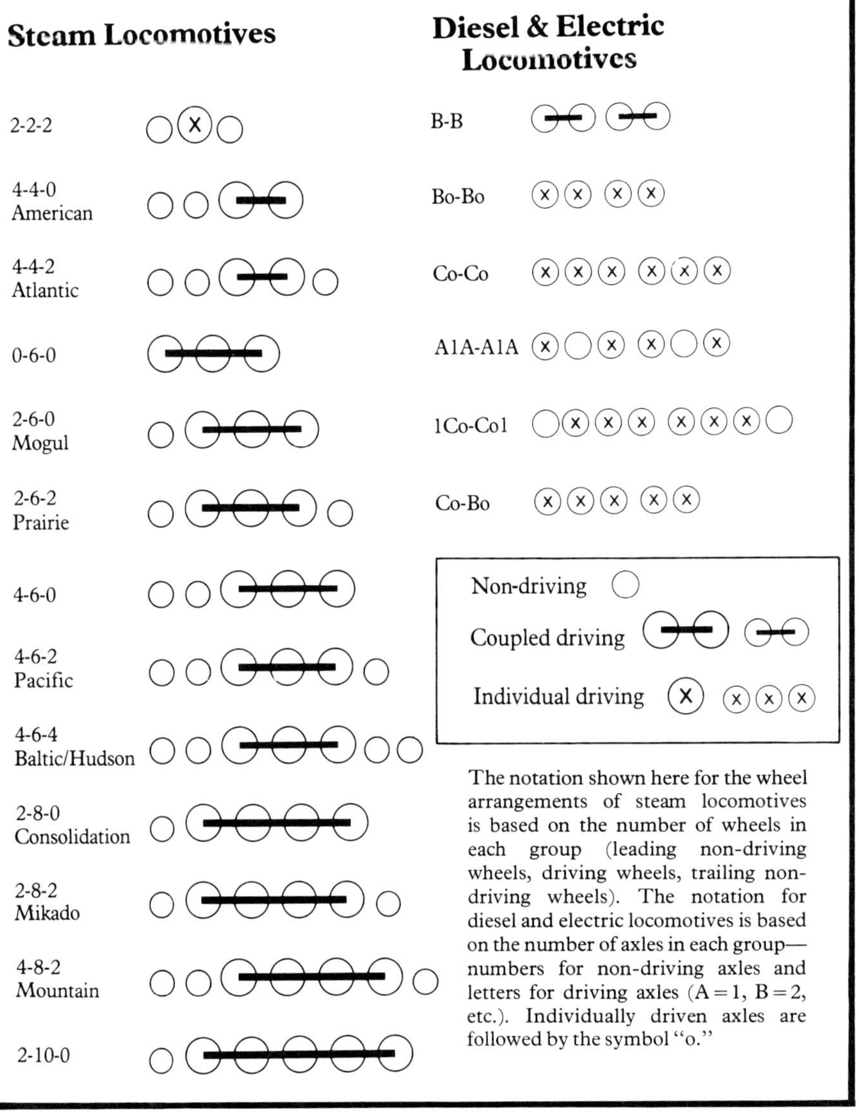

**Steam Locomotives**

- 2-2-2
- 4-4-0 American
- 4-4-2 Atlantic
- 0-6-0
- 2-6-0 Mogul
- 2-6-2 Prairie
- 4-6-0
- 4-6-2 Pacific
- 4-6-4 Baltic/Hudson
- 2-8-0 Consolidation
- 2-8-2 Mikado
- 4-8-2 Mountain
- 2-10-0

**Diesel & Electric Locomotives**

- B-B
- Bo-Bo
- Co-Co
- A1A-A1A
- 1Co-Co1
- Co-Bo

Non-driving ○
Coupled driving
Individual driving ⓧ

The notation shown here for the wheel arrangements of steam locomotives is based on the number of wheels in each group (leading non-driving wheels, driving wheels, trailing non-driving wheels). The notation for diesel and electric locomotives is based on the number of axles in each group—numbers for non-driving axles and letters for driving axles (A = 1, B = 2, etc.). Individually driven axles are followed by the symbol "o."

# Index

**A**
Abt, Roman, 26
Africa, railroads in, 6
agricultural-produce car, 12
Amtrak (National Railroad Passenger Corporation), 9
articulated tram, 29
Asia, railroads in, 7
atmospheric railroad, 22
Australia, 6, 7
automatic-control train, 28

**B**
Baltimore and Ohio Railroad, 8
Bay Area Rapid Transit, 28
Berlin Trade Exhibition, 20
*Best Friend*, 19
Blenkinsop, John, 26
*Blue Train*, 10
bridge, railroad, 16
building of railroads, 5-6, 17

**C**
*California Zephyr*, 11
cantilever bridge, 16
carriages, railroad, 4
"chairs," 14
commuter trains, 9
container traffic, 13
control yard, 12
cuttings, 17

**D**
diesel locomotive, 18, 19, 20-21
diesel-electric locomotive, 20, 21
dining car, 8
double car, 12
double engine, 22
driver's cab, 25

**E**
Eiger Mountain, 27
electric locomotive, 18, 20
embankment, 17
engineering, 16
England, railroads in, 4, 6, 23, 27, 28, 31

**F**
flange, 4, 15
Forth Bridge, 16
France, railroads in, 9, 10, 28
freight train, 5, 12, 13

**G**
Gooch steam engine, 28
Grand Central Station, 17

**H**
highest railroad, 7
Hikari train, 11
hovercraft, British tracked, 23
Howey, Captain S.E.P., 27
"hump," 12

**I**
Indian Pacific Express, 7
inter-city trains, 11
"iron horse," 6

**J**
Japan, railroads in, 7, 11, 16
Jungfrau railroad, 27

**L**
Lartique monorail, 22
Lightning train, 7, 11
liquid gas tanker, 12
*Locomotion*, 30
locomotive, 4, 6, 7; development of, 19; diesel, 18, 19, 20-21; diesel-electric, 20, 21; electric, 18, 20-21; steam, 18, 19, 20, 25, 31
Loetschberg Tunnel, 16

**M**
*Mallard*, 31
Messerschmitt-Bolkow-Blohm (MBB), 23
Métro, 28
model railroad, 31
monorail, 28
Moscow underground, 29
mountain railroad, 26, 27

**N**
narrow gauge track, 26
National Railroad Passenger Corporation (Amtrak), 9
National Railway Museum (England), 31
North America, railroads in, 4, 5

**O**
*Orient Express*, 10

**P**
"Pandrol" clip, 14
passenger train, 5, 6, 8, 9
permanent way, 14
Pilatus, Mt., 26
preserved railroads, 30
Pullman Car Company, 10

**R**
rack and pinon system, 26, 27
rail, 14, 15
*Rheingold*, 11
*Rocket*, 19
Romney, Hythe, and Dymchurch Railroad, 27
Russia, railroads in, 10, 29
*Russia, The*, 10

**S**
safety, 24
screw-driven train, 22
semaphore signal, 24
shuttle train, 28
signal box, 24
signals, 24
Simplon tunnel, 16
sleeping compartment, 8
South America, railroads in, 7
standard gauge track, 7, 26
station, railroad, 17
steam locomotive, 18, 19
Stephenson, George, 30
Stephenson, Robert, 19
Swaziland Railroad, 6
switches, 15
Switzerland, railroads in, 16, 17, 26, 27

**T**
tie, 14, 24
track, 4; laying of, 14; gauges of, 7, 15, 16, 26
trams, 29
Trans-Europ-Express (TEE), 11
Trans-Europ-Express-Marchandise, 13
Trans-Siberian route, 10
transcontinental trains, 10
transit train, 28
Trevithick, Richard, 19
tunnels, railroad, 16

**U**
underground railroad, 28-29
United States, railroads in, 6, 8, 9, 11, 30

**V**
vehicle transporter, 12
viaduct, 17

**W**
wagon, 4, 12
wagonway, 4
wheel arrangements, 34

Lerner Publications Company
241 First Avenue North, Minneapolis, Minnesota 55401